UNDERSTANDING
RADIATION

I0472951

A COMMON SENSE
APPROACH

**The basic facts about radiation
which every citizen should know**

**by
Blaine N. Howard
Radiological Physicist**

Published: January 2012

Cover Photo: public domain –
Susquehanna steam electric station

Copyright © 2012 Blaine N. Howard
ISBN-13: 978-1-4679-9423-1
ISBN-10: 1-4679-9423-5

I would like to thank my wife, Kathleen, for allowing me to spend so much time in writing and rewriting my book.

I also want to thank my son, Carl, for his help in preparing the material for publication and taking responsibility for actually getting the book published.

I also want to thank all my friends who have encouraged me by saying, "You ought to write a book" as I tried to explain some of the facts of radiation to them.

Finally, I want to thank those who have read my attempts and offered much helpful advice, especially Craig Jones, Bruce Howard, and Kathy Williams.

Table of Contents

Introduction

I have observed that many political decisions involving radiation and radioactive materials have been made by persons with very limited knowledge about radiation and its effects on health. I have also observed many articles published by the news media which reveal a serious misunderstanding of basic facts about radiation and its effects. For this reason, I wanted to make some of the basic facts about radiation and radiation effects readily available to those in political offices which require decisions about radiation and nuclear power. In doing this, I also want this information to be available to members of the news media who write about such matters and to the general public who may be overly concerned about harm from nuclear radiation.

Have you ever read a newspaper article claiming that thousands of Utahns have suffered from fallout or that anyone living within one mile of a radioactive waste transport route was in danger? Have you heard about Congress considering banning the importation of nuclear waste? Have you heard about the Utah legislature's opposition to the "Divine Strake" test proposed by the Defense Threat Reduction Agency and the public outcry against it? [Divine Strake was a proposed chemical explosion on the Nevada Test Site]

Or have you heard rumors that radiation from the Fukushima nuclear reactor leaks in Japan may be dangerous to people in the United States? Did the radiation from the Chernobyl accident cause any health problems in the United States?

Do any of these questions cause you to wonder what to believe and whom to trust? If so, read about some of the facts which should remove much of the mystery.

I propose to give you basic facts which you can understand and sources for those facts which you can trust. Some of these facts include:

- How much radiation we get from nature.
- How much radiation we get from a nuclear waste site.
- How much radiation it takes to cause radiation sickness and death.
- How much radiation it takes to increase the risk of cancer.
- How many people die from cancer normally.
- Two basic theories about low dose radiation effects.

In order to tell you some of the things which have concerned me for many years about the lack of understanding of the basic principles of radiation and its effects on our health, I first have to explain some terms and ideas which are well known to health physicists and other radiation scientists. I will try to explain them in terms which are easily understood by those unacquainted with scientific terms.

Radiation

Radiation is anything that radiates or that is emitted outward. Nuclear radiation is emitted from the nucleus of an atom. There are 3 types of nuclear radiation: alpha, beta, and gamma. They are listed in their ability to penetrate. Alpha can be completely stopped by a sheet of paper. Beta can be completely stopped by about 2 inches of tissue. Gamma radiation is not completely stopped by any thickness of shield. If we add enough shielding to reduce the intensity to ½ its original value, this is called a half value layer. Adding an additional half value layer will reduce it by ½ of the remaining radiation to ¼ of the original amount. Continuing to add half value layers will keep reducing the radiation by halves as 1/4,1/8,1/16 etc. but, theoretically, it would never reach zero.

Radioactive Decay

When an atom is radioactive it is not stable, but will change into a different kind of atom. We call this radioactive decay. An example is radium-226 decaying to radon-222.

When an atom decays, it sends out radiation. This is called "nuclear radiation" because it comes from the nucleus of the atom.

Radiation coming from outside our bodies is called "external radiation" and radiation coming from within our bodies is called "internal radiation".

I will be discussing radiation coming from our environment (external radiation) and affecting our whole bodies. This means that I will be discussing gamma radiation as well as x-rays which are very similar to gamma rays.

Half-lives

The amount of time for ½ of the atoms in a radioactive sample to decay is called its half life. After a second half life, only ¼ of the original atoms remain. After 10 half-lives, about one one-thousandth of the original atoms will remain.

Each "isotope" has its own characteristic half-life. For example, cesium-137 has a half-life of 30 years and is the longest half-life fission product. That is why scientists say that reactor waste will not be dangerous after 300 years (10 half-lives of cesium-137). Fission is the splitting apart of atoms and is the process by which nuclear reactors get their energy.

Detecting and Measuring Radiation

An important thing to know about radiation is that it cannot be detected by our senses. That is, we cannot see or feel radiation. We must have special instruments to detect and measure radiation. Since you probably do not have these special instruments, I will give you some of the results of measurements made and reported by scientists.

In order to measure radiation, we need to have units to express the amount of radiation we measure, just as we need units of weight to measure how much you weigh. The units of weight are pounds. The units of radiation are rem. To understand what a rem is you need to know how many rem it takes to cause an effect on your health.

If 100 rem are received within a few hours, it may cause "radiation sickness". This is a very large "dose" of radiation. We speak of receiving a "dose" of radiation just as we speak of taking a "dose" of medicine. If the 100 rem dose is not received in a short time but is received over a long time like a year, the result will be much different since the body will have time to recover from the harm. Since a rem is a very large dose of radiation, we would like to measure radiation in smaller units. To do this scientists just add a prefix to the unit to make it smaller. The prefix "milli" means one-thousandth so a millirem would be one-thousandth of a rem. In other words, it takes 1,000 millirem (abbreviated mrem) to make one rem.

You may be wondering if a mrem can have an effect on your health. Such an effect, if it exists, is too small to measure. But, if it exists, it may have an effect on your chance of getting cancer. There are theories about this possibility. One theory claims that it increases the risk of cancer and another theory claims that it decreases the risk of cancer. The calculated effect of even 5,000 mrem is still too small to see. Since the effect, if it exists, is too small to see, it doesn't make much difference to our health which theory is true. So, if you have been told that you should worry about health effects of a few mrem of radiation, you can stop worrying. The health effect of the worry is greater than that of the radiation.

There is another unit of radiation you should be aware of. That is the "Sievert". The rem is the unit still used most frequently in the United States while the Sievert is an international unit. A Sievert is 100 times larger than a rem, so a Sievert is about a radiation sickness dose.

Scientists in the United States are trying to convert to using the international unit so I will try to express my radiation doses in both units. A millisievert is abbreviated mSv.

If you are a private citizen, I hope that the understanding you receive from this short booklet will help you make good decisions about your personal radiation exposures such as x-rays and nuclear medical procedures.

If you have connections with the news media, I hope that you will use your connections to make sure all future media coverage of radiation related matters are in agreement with fact.

If you are in the political arena, I hope that you will influence future decisions and legislation to conform to facts and not rumors.

If any of the material seems unclear or if you have any doubts about the veracity of the statements, I welcome the opportunity to clarify them. You can send questions to:

> Blaine N. Howard
> blaine.howard@understanding-radiation.com

Chapter 1: **How Much?**

Every citizen should know that radiation is not new. It did not begin with the atomic age or the discovery of x-rays. Our ancestors lived with it. The following facts are important to know in order to make any reasonable judgments about radiation or radioactive materials.

The Facts

- **We receive more radiation every year from nature than most people realize – more than one hundred times what we could get from a properly functioning nuclear facility.**

- **Radiation from man-made sources to the general population is trivial compared to that from nature. A few individuals may receive significant exposures from rare accidents but the doses to the general population, even for these accidents, is small.**

The Supporting Evidence

I will now try to explain these facts in detail so that you will be able to trust them.

The reason most people do not realize how much radiation we receive all the time from nature is that we cannot detect radiation with our senses. We could be in a radiation-free environment or in a high-radiation field and never know the difference. The

1

only way we can tell how much radiation we are receiving is by measuring it with a special meter designed for that purpose. Very few people have access to a radiation meter so I will give you the results of measurements which have been made. To express this information, I will need to use radiation dose units.

Units of Radiation

As mentioned in the introduction, there are two different sets of units used to express the doses.

In the United States, most radiation doses are expressed in units of "millirem" abbreviated mrem.

In international units, radiation doses are expressed in units of "millisieverts", abbreviated mSv.

You can find a brief definition of these units in the introduction, but I want to emphasize that 1 mSv is equal to 100 mrem. I will try to express doses of radiation in both units to make them understandable to everyone.

I have chosen to discuss the background radiation in Utah because I have this information readily available. The radiation in other states should have similar variations but the average will be unique for each state. The average background radiation for the United States as a whole, is about one half that of Utah.

Note: The Utah background radiation which I quote is the measured gamma radiation and does not include radon, which emits alpha radiation.

Utah Background Radiation

The ionizing radiation which we receive from nature is called "natural background radiation" or simply "background radiation". In 1979, scientists from the Environmental Measurements Laboratory in New York City (part of the US Department of Energy) came to Utah and spent much of the summer measuring the background radiation at many locations. They measured the radiation coming from the radioactive materials in the environment as well as that which comes from outer space called "cosmic radiation".

I have a copy of the report of these measurements which is titled "The Natural Radiation Background in Utah – Preliminary Report on Radionuclides in Soils in Populated Areas". I will take information from this report to present a valid picture of the Utah background radiation. Figure 1 presents the average reading for each selected city along with both the highest and lowest readings for that city. These readings represent the total of both the gamma radiation from the environment and cosmic radiation from outer space. Units are mrem per year.

Included in these measurements is the amount of gamma radiation from cesium-137 (Cs-137) in the soil. This is the only gamma emitter remaining from the Nevada Test Site (now the Nevada National Security Site) fallout. Since Cs-137 has a half-life of 30 years, the amount remaining in 2011 will be about one half of what it was in the 1979 measurements. I have not listed this measurement separately but the values ranged from 0 to 4 mrem per year and averaged just less than 2 mrem per year. So the 2011 values would average less than 1 mrem per year.

Notice the numbers on the left side of the chart. This is the amount of radiation which is received at these locations and is expressed in units of millirem per year or simply mrem/yr. By comparing the numbers, you can get a "feel" for how large the unit is.

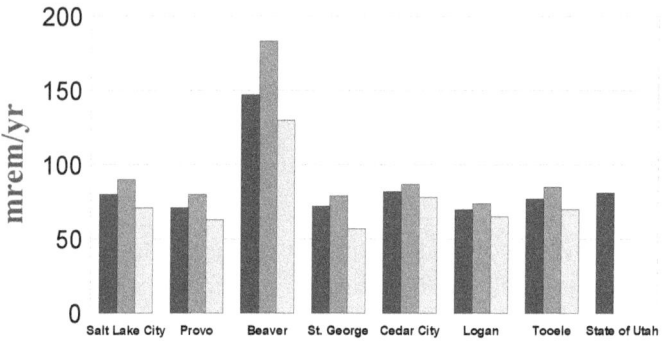

Figure 1. Background Radiation in Utah cities including cosmic rays.

As you study Fig. 1, you should notice that the amount of radiation varies from city to city. Beaver, Utah, has a very high background compared to other cities in this survey. This is because of the higher amounts of naturally occurring radioactive material in the soils. It has an average of 147 mrem/yr but the readings ranged from 130 mrem/yr to 183 mrem/yr. This means that residents in one part of Beaver are getting 53 mrem/yr more than residents on another part of the city. In most cities, this difference is much less. In Salt Lake City, Utah, the difference was measured to be 19 mrem/yr.

The amount of radiation one receives from the natural background depends not only on where in the state but also where in the city one lives.

Here is some additional information about the effects of short-term radiation doses.

- 25,000 mrem (250 mSv) is enough to cause a noticeable change in a person's blood count.

- 100,000 mrem (1,000 mSv), when received in a short time, may cause some radiation sickness with nausea and vomiting.

- 500,000 mrem (5,000 mSv), when received in a short time, may cause death in some people within about a month.

Scientists call the amount of radiation received by an individual a "dose". Although very high doses of radiation can be very dangerous, the effect of low doses may be quite different. I will be discussing doses of radiation much lower than those which could produce a noticeable short term effect.

As you study Fig. 1, you may think that these background radiation rates are very high and something to be worried about. However, there are some other areas of the world with much higher background radiation rates. These areas do not have an increased cancer rate associated with the increased background radiation. The following information comes from the Health Research Foundation at Kyoto, Japan. The chart gives both the average and the highest reading at each location.

World Wide Background Radiation		
Area	Average Dose mrem/yr	Maximum Dose mrem/yr
Ramsar, Iran	1,020	26,000
Guarapari, Brazil	550	3,500
Kerala, India	380	3,500
Yangjiang, China	351	540
Hong Kong, China	67	100
France	60	220

Germany	48	380
Japan	43	126
USA	40	88

The average values for the highest background cities are shown in Fig. 2 compared to the average readings for some Utah cities.

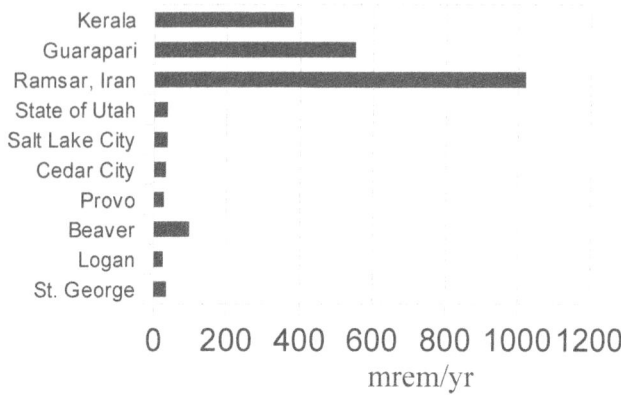

Figure 2 World Background Compared to Utah Background

This should give you enough information to establish firmly the first fact I listed. That is:

"We receive more radiation every year from nature than most people realize." It should also let you know that what we receive in Utah is only a small fraction of the maximum background levels in the world.

Radiation From Man-Made Sources

A. X-rays

The most familiar source of man-made radiation is the x-ray machine. While x-rays are not nuclear radiation, they are similar to gamma rays. Nearly everyone has had some type of x-ray procedure and most people have had many of these.

It is important to understand the difference between x-ray exposures and nuclear radiation exposures. While nuclear radiation usually exposes the whole body, x-ray examinations are confined to only a part of the body. Thus, the average dose to the body is only a fraction of the dose to the region of interest.

The following chart gives some of the typical x-ray doses to the part of the body being examined. This chart comes from the American Nuclear Society web site dated July 22, 2011.

X-ray	Dose	
	mrem	mSv
Dental bitewing	0.5	0.005
Chest	10	0.1
Cervical spine	20	0.2
Mammogram	42	0.42
Lumbar spine	600	6.0
Upper GI series	600	6.0
Abdomen (kidney/bladder)	700	7.0
Barium enema	800	8.0
CT scan whole body	1,000	10.0
CT scan heart	2,000	20.0

B. Nuclear Sources

I will now discuss the amounts the public could receive from man-made nuclear radiation sources. The first source I wish to discuss is a nuclear reactor. Reactors are used to power ships such as submarines and aircraft carriers and to supply electrical power to cities. These facilities are regulated very strictly by federal agencies. The Nuclear Regulatory Commission requires that radiation to the public be limited to 25 mrem per year at the fence line to the facility property. That means that, for a person to receive 25 mrem from the facility, he would have to stay at the fence line for 24 hours per day for 365 consecutive days. I would guess that a person who chose to expose himself to radiation would not be willing to camp out at the fence line of a reactor for more than a week—probably much less. Since there are 52 weeks in a year, the dose for one week would be 25 divided by 52 or 0.48 mrem maximum. A person will not be exposed against his will.

This same limit would be in effect at a nuclear waste storage site. That is less than one half of a mrem while camping out for a solid week at the property line. And this is if the facility is allowing the maximum permitted radiation at the fence line. It usually is operating below this limit.

The most radiation which any other facility may allow to reach the public is 100 mrem per year or 4 times the limit for a nuclear reactor.

Consider the dose to someone who goes onto a site as a visitor. How much radiation could a visitor to the site receive? I looked at actual personnel monitoring records for the EnergySolutions site at

Clive, Utah. I refer to a copy of the 2005 Personnel Exposure Report from EnergySolutions. I was given this report as an example of exposures during a year in which higher than normal quantities of waste were being handled. Here is a summary of that report.

Total number of dosimeter badges	536
Number of badges with "zero" dose	418
Average dose for the other 118 badges	17 mrem/yr
Highest badge reading	414 mrem/yr

Now let's make a little calculation to estimate how much exposure a visitor could expect to receive from a visit to EnergySolutions. Most likely, he would not be in any radiation areas at a higher dose rate than the average of the non-zero badges or 17 mrem/yr. But he would only be there for a short time, certainly less than 8 hours. If we divide 8 hours by 2,000 hours (40 hours per week for 50 weeks) and multiply by 17 mrem we get 0.068 mrem. A badge with that much radiation would read "zero" since all readings less than 0.5 are reported as "zero". Now let us look again at Fig. 1 with an added column using 1 mrem (15 times what we calculated) for EnergySolutions. This is shown as Figure 3.

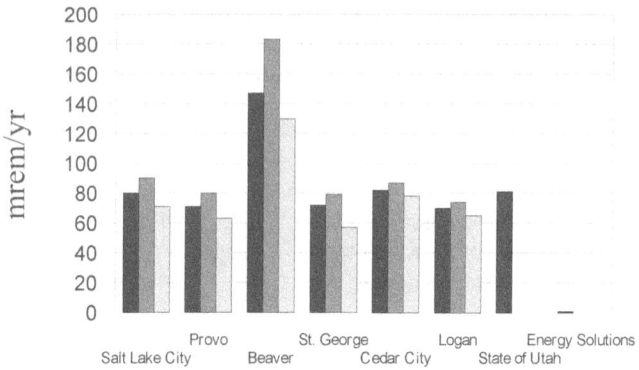

Figure 3 Utah Background Radiation Compared to
EnergySolutions

From Figure 3, it should be obvious that the
exposure a visitor to EnergySolutions might expect is
much less than that received all the time from
background radiation.   To make it easier to see,
multiply the 0.068 mrem by 1,000 and get 68 mrem.
Thus 1,000 times what a visitor might expect to
receive from a visit to the EnergySolutions site is still
less than the average background radiation in Utah.

This should be enough evidence to establish the
second fact I stated. That is:

**"Radiation from man made sources to the
general population is trivial compared to that
from nature."** To make it more obvious, consider
that the number of citizens who visit a radiation site
or even come within a mile of it is very, very small
compared to the total population.

I appeal to your common sense now to judge
whether anyone should fear being harmed by
radiation from the EnergySolutions site—or any other

nuclear facility. We assume here that natural background radiation is not harmful to our health. This fact will be established in chapter 3.

Nuclear Accidents

Sometimes, in a nuclear accident, a few people receive significant doses. Consider the Chernobyl accident. High doses of radiation to rescue workers and employees of the plant caused 28 deaths among 134 patients who were confirmed to have radiation sickness. ("The Chernobyl Disaster and How It Has Been Understood" by Zbigniew Jaworowski)

Radiation to the public was elevated. In Poland, the initial dose rate was 2.6 mSv per year (260 mrem per year). ("Lessons of Chernobyl" by Zbigniew Jaworowski)

The Russians evacuated 336,000 people from the regions of the former Soviet Union where the radiation dose from Chernobyl fallout was about twice the natural dose. These are small compared to the 10.2 mSv per year (1,020 mrem per year) background radiation in Ramsar, Iran. ("The Chernobyl Disaster and How It Has Been Understood" by Zbigniew Jaworowski)

Although some people reccived lethal doses of radiation, the radiation to the public was still within the range of normal background radiation.

Radioactivity in the Human Body

Another fact which few people realize is that our own bodies contain a significant amount of radioactive material. We receive about 20 mrem/yr (0.2 mSv/yr) from the Potassium-40 in our bodies.

Each second, over 4,000 potassium atoms decay within your body. Counting all the radioactive istopes in your body, over 8,000 atoms decay each second. See the following chart for more details. The activity is the number of atoms which decay each second.

Activity Within the Body
in disintegrations per second

Isotope	Activity (dis/sec)
Potassium 40	4,340
Carbon 14	3,080
Rubidium 87	600
Lead 210	15
Tritium	7
Uranium 238	5
Radium 228	5
Radium 226	3
Total	8,055

Summary

The amount of radiation you could receive from a nuclear facility such as EnergySolutions is so small (less than 1 mrem) that it is dwarfed by the amount of radiation you receive from diagnostic x-rays or even the amount of radiation you receive from the radioactive materials in your own body.

The amount of radiation you receive from nature is much much larger than this—over 100 times larger. And this "background radiation" is not harmful to your health.

To believe that limiting the amount of radioactive waste coming to EnergySolutions or keeping them from receiving "hotter" waste will have any influence on the amount of radiation received by the citizens of Utah is completely without reason.

Chapter 2: **The Health Effects of Radiation**

Rumors

You have been told that:

- "All radiation is harmful." and

- "Every bit of radiation you receive adds to your chance of getting cancer."

These rumors are not true. To understand why they are not true you must understand what effects radiation can have on your health.

There are two kinds of effects to consider. They are first, short term effects which are produced by very large doses of whole body radiation and second, long term effects which are produced by smaller doses of radiation.

Short Term Effects

The short term effects which include sickness and death require a tremendous amount of radiation and very few people have experienced these effects. These effects appear within a few hours to a few weeks after exposure. Extremely large doses of radiation which are given to tumors in radiation therapy do not cause these effects since it is not whole body radiation.

The Facts

The facts which I wish to state about these effects are:

- **Radiation exposures of about 100,000 mrem (1,000 mSv) and greater cause radiation sickness.**

- **Radiation exposures of about 400,000 mrem (4,000 mSv) and greater may cause radiation death.**

The Supporting Evidence

To produce these effects, the radiation must be received in a short time (probably a few hours).

Radiation sickness will be evident shortly after the exposure—usually within hours. Radiation caused death may occur within a short time for extremely high doses but usually within two months. There are also other short term effects including radiation burns.

These radiation effects have been well established. The only question about these facts is the accuracy of the doses which produce the effects. That is the reason I have listed them as "about". I will not try to establish accurate numbers because there are none. There is a lot of variability in people and so there should be a range of numbers which could apply. The important thing about these effects is that there is no question about their existence. You can find a chart of the approximate doses and effects at various websites on the Internet.

Long Term Effects

Whereas short term effects are seen within hours of exposure, long term effects may take over twenty years to develop. The most commonly feared long term effect of radiation is cancer. The important facts for you to understand about radiation and cancer are the following:

The Facts

- **There is no conclusive evidence that doses less than 5,000 mrem (50 mSv) cause an increase in the risk of cancer.**

- **Estimates for cancer deaths from doses lower than 5,000 mrem (50 mSv) are just calculations made using a theory which has not been proven and, thus, is an unreliable theory.**

Note:

Genetic effects: Although genetic effects could span generations, no significant genetic effects have been found at these low dosages.

Monsters: Science fiction movies and stories have suggested that mutations caused by radiation can produce monsters. Such fictional Radiation-induced mutations have not been observed.

The Supporting Evidence

Now, let me show you that doses less than 5,000 mrem (50 mSv) have not shown conclusive evidence of increasing the risk of cancer.

Because of the sparsity of good data for the lower doses, scientists have not been able to agree about the radiation effects of doses less than 40,000 mrem (400 mSv). A summary of these views is stated by the General Accounting Office as:

> *"According to a consensus of scientists, there is a lack of conclusive evidence of low level radiation effects below total exposures of about 5,000 to 10,000 millirem."*
> GAO/RCED-00-152, **Radiation Standards**.
> Page 10. June 2000.

This statement is in good agreement with a Position Statement from the Health Physics Society which includes the following information:

> *"There is substantial and convincing scientific evidence for health risks following high-dose exposures. However, below 5–10 rem (which includes occupational and environmental exposures), risks of health effects are either too small to be observed or are nonexistent."*
> "Radiation Risk in Perspective"PS010-2 , revised 2010

(Note: 5 to 10 rem is equal to 5,000 to 10,000 millirem and 50 to 100 mSv.)

The Health Physics Society is a professional organization whose mission is excellence in the science and practice of radiation safety. Of all the professional societies, the Health Physics Society is the one specializing in the effects of ionizing radiation and should be the first choice for such information. The Society has approximately 5,000 members in 48 countries.

A similar statement was made by the National Academy of Sciences committee on the Biological Effect of Ionizing Radiation or BEIR committee. They state that:

> *"With few exceptions, however, [cancer] effects have been observed only at relatively high doses and high dose rates. Studies of populations, chronically exposed to low level radiation, such as those residing in regions of elevated natural background radiation [10 - 100 times average US levels], have not shown consistent or conclusive evidence of an associated increase in the risk of cancer."*
>
> Health Effects of Low Levels of Ionizing Radiation. Committee on the Biological Effects of Ionizing Radiation (BEIR V). Page 5. National Academy of Sciences, 1990.

The highest level of natural background radiation was measured in Ramsar, Iran where the average is 1,020 mrem (10.2 mSv) per year. The maximum rate in Ramsar is 26,000 mrem (260 mSv) per year.

While the BEIR committee did not give a precise range of dose rates, the background radiation levels they refer to are in basic agreement with the other two quotes.

I quote these recognized authorities to establish that **there is no conclusive evidence that doses less than 50 mSv (5,000 mrem) cause an increase in the risk of cancer.**

Other Views

However, contrary to these statements, there are many who claim that low doses of radiation do increase the risk of cancer. They base this on the

Linear No Threshold (LNT) model. The LNT model claims that the risk of cancer is proportional to the dose no matter how small the dose. It will be discussed thoroughly in chapter 3. They assume the Linear No Threshold (LNT) model of radiation effects to be valid for extremely low doses. In fact, they assume that this model is a true representation of what actually happens and is valid all the way to zero dose. What these people fail to tell you is that the calculated increased risk is too small to be detected— even for 50 mSv (5,000 mrem).

The increased risk of dying from cancer resulting from a dose of 50 mSv (5,000 mrem) predicted by the LNT is 1 per 1,000 whereas the normal chance of dying from cancer is 230 per 1,000. Since the chance of dying from cancer varies from year to year by much more than this calculated amount, the calculated increased risk is too small to be detected. (See chapter 4 for more information about this)

Let me emphasize that this LNT model has serious limitations and does not hold true for doses less than 50 mSv. The International Commission on Radiological Protection (ICRP) makes the following statement about the LNT model.

"… The risk factors for the lower dose ranges that are important for radiological protection as well as the concept of effective dose, are based on the linear-no-threshold model (LNT model). This model is an assumption which has not been scientifically validated.

ICRP 103 (2007)(Page 317)

It is important for you to realize that <u>all predictions</u> of deaths resulting from the exposure of large numbers of people to small doses of radiation are just <u>calculations</u> using the LNT model and, since the LNT is only an assumption, they are <u>not reliable</u>.

Thus the thousands of excess cancer deaths which were predicted to result from the Chernobyl radiations are not valid.

Also, the thousands of excess cancer deaths which have been predicted to result from the fallout from the nuclear weapons tests are not valid.

These are only calculated numbers based on a model which has not been proven.

There is good evidence that the fallout from nuclear weapons tests resulted in a decrease in the cancer death rate.

1. Dr. Charles R. Smart, who founded the Utah Cancer Registry, testified in court as reported in the New York Times November 8, 1982 , referring to southern Utah,

> **"But from 1966 to the present time there is no excess," he said. "In fact, there is the opposite."**

"The opposite" means <u>less cancer</u> in southern Utah.

2. S. G. Machado, C. E. Land and F. W. McKay reported that

> *"Mortality from all cancer sites combined was significantly lower in southwestern Utah*

than in the remainder of the state, even after adjustment for the higher proportion of (lower risk) Mormons in southwestern Utah ." See "Cancer Mortality and Radioactive Fallout in Southwestern Utah" *Am. J. Epidemiol. (1987) 125(1): 44-61*

3. Dr. Ray Lloyd, of the University of Utah analyzed leukemia data for Washington County, hoping unsuccessfully to find an excess. He stated :

"If essentially no leukemias were induced among the Washington County population by NTS fallout (see Table 1), then virtually no other cancers were induced." See Health Physics, June 1997, Volume 72, Number 6 pages 938 through 940

4. More recently, the Deseret News on April 13, 2011 reported :

"Information that was compiled by the National Cancer Institute showed for a 30-year period, there was an 8 percent higher rate of radiation-associated cancers throughout the rest of the state than in counties where residents receive compensation."

This means that, for this 30 year period, residents in counties receiving compensation had an 8 percent <u>lower</u> cancer rate than the rest of the state.

For more information see a book by Dr. Daniel Miles titled "The Phantom Fallout-Induced Cancer Epidemic in Southwestern Utah: Downwinders

Deluded and Waiting to Die", BookSurge Publishing 2009.

Chapter 3: **Models and Theories**

It will be helpful, in understanding the risks of radiation, to know something about the theories and models which scientists use in studying their data.

There are two main theories or models of how radiation-caused cancer varies with the amount of radiation. They are the Linear No Threshold or LNT model and the Radiation Hormesis model.

The **LNT model** claims that all radiation increases the risk of cancer proportional to the dose for all doses even to zero dose.

The **Radiation Hormesis model** claims that doses less than 40,000 mrem (400 mSv) reduce the risk of cancer by triggering the body's defense mechanisms.

The public deserves to know that the LNT model has never been proven. All calculations of predicted deaths for doses less than 5,000 mrem using the LNT are only calculated numbers and are unreliable.

The public also deserves to know that there is another model, the Radiation Hormesis model. Although this model has not been accepted by regulatory authorities and has not been proven, it agrees with existing data and therefore offers a more realistic point of view than the LNT model.

In this chapter I will explain the following facts:

The Facts

- **The LNT model explains the relationship between radiation dose and excess cancer** and corresponds to the data for high doses, but does not match the data for doses less than about 40,000 mrem (400 mSv).

- The Radiation Hormesis model explains the relationship between radiation dose and cancer for doses less than 40,000 mrem (400 mSv) and is more successful for these doses than the LNT model. It claims that, for these low doses, the effect of radiation is to reduce cancer.

The Supporting Information

The LNT Model

Let me describe just what the LNT model is. First we construct a graph of the increased cancer mortality (death rate) as a function of the radiation dose. We draw a vertical line and label it "**Excess Cancer Mortality**". Then we draw a horizontal line from the lower end of the vertical line and label it "**Radiation Dose**". Then we would plot the data as points on this graph if we had such data.

To illustrate this, let me plot some dots on the theoretical graph as shown below (Figure 3a). On a real graph of real data, the vertical and horizontal lines would be marked with calibration marks and each point would represent a single bit of data which is the amount of excess cancer mortality for the given radiation dose as found in a scientific study.

While there is good information about the radiation effects at high doses, there is little good data for the low doses. The circle indicates the area of uncertainty. The solid line represents the expectations of the LNT model. The dashed line represents the radiation hormesis expectation.

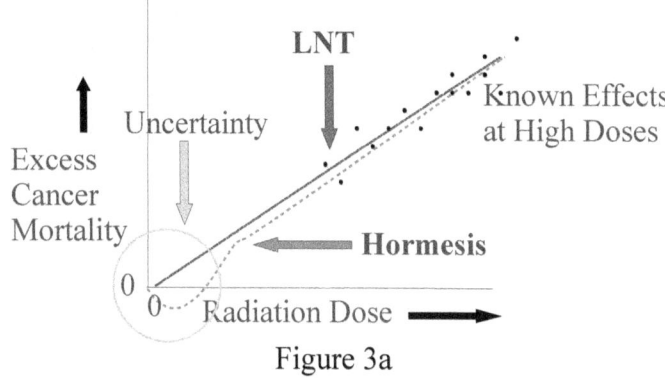

Figure 3a

Cancer occurs commonly in normal populations but study of the atomic bomb survivors showed an increased number of cancers. To see the relationship of the cancer increase to the radiation dose it is necessary to know both the cancer rate and the radiation dose. From what is known about the radiations coming from an atomic bomb and by noting where each individual was at the time of the explosion, we can get a good idea of how much radiation each individual received from the blast. We can then divide the survivors into groups which received similar amounts of radiation.

Cancers which are caused by radiation take a long time to develop but, in over fifty years, there have been enough cancers to provide some good information. If we plot the excess cancer rate against the dose for each group of survivors we can see how the increased cancer rate relates to the dose. The following plot (Figure 3b), provided by Dr. Bernard L. Cohen, shows some of the actual data from the atomic bomb survivor study.

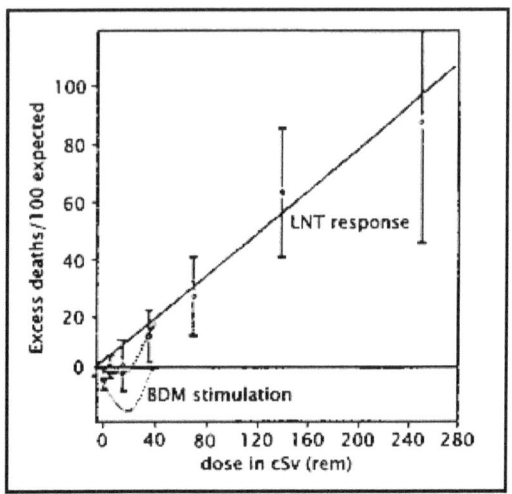

Fig. 3b. Excess deaths from solid tumors per 100 "expected" among Japanese A-bomb survivors (1950-1990) vs Dose.(11) Error bars are 95% confidence limits. Lines represent a proposed analysis into an LNT component (solid line) plus a contribution at low dose from biological defense mechanisms (BDM --- small dot line) to give a resultant behavior at low dose shown by large dot line, merging into the LNT line above 50 cSv. [500 mSv]

Data from *Pierce D.A. et al, Studies of the mortality of atomic bomb survivors, Report 12, Part 1, Cancer 1950—90, Radiation Research, vol. 146, p1—27, 1996.*

Let me explain the data points. [Don't be confused that the dose is given in cSv. There are 10 mSv in one cSv so just multiply by 10 and you have the dose in mSv.] The dots are the actual data from the study. These points are not precise values but only estimates of the true value. The vertical lines through each point are 95% error bars – which means that there is a 95% chance that the true value is somewhere in this range. There is also a 5% chance that the true value lies outside the error bars.

A scientist tries to draw a line through the data which lies within all of the error bars. In this case, the lowest dose point has error bars which do not include the proposed line which represents the LNT response. The point for the lowest dose is actually below the zero excess deaths line which means that there were less cancer deaths than would normally be expected.

It should be obvious, even to a non scientist, that the lower dose points seem to get farther and farther away from the LNT line. What Dr. Cohen proposes is that an additional effect at these low doses can bring the line back into agreement with the error bars. This effect is from the body's biological defense mechanisms acting to protect the body from cancer. When this correction, represented by the lower dotted curve, is added to the LNT line, the result is the upper dotted curve and conforms well to the data. This protection from cancer is called "radiation hormesis".

Radiation Hormesis

Now, let us evaluate what we have just observed.

- It is true that the lower dose points on this plot are not in agreement with the LNT model.

- It is also true that Dr. Cohen's addition of a hormesis correction brings the curve into agreement with these data points. The lowest doses result in fewer cancer deaths than normal.

Common sense would tell us that, if what Dr. Cohen is saying is true, there should be some evidence of this effect (reduced cancer rates) among people who get small doses of radiation.

Where can we find a group of people who regularly receive small doses of radiation in addition to background radiation? I think that common sense should tell us that workers in the nuclear energy business are the most likely candidates. How is their cancer death rate effected?

The LNT model would have us expect to see elevated cancer death rates among this group of workers. The Radiation Hormesis model would have us expect to see lower cancer death rates among this group of people. Again, appealing to your common sense, you should be willing to accept the implications of the actual cancer death rates among these workers.

Fortunately we have information about the cancer death rates for these workers. In 2004, a report was published in the AMERICAN JOURNAL OF INDUSTRIAL MEDICINE entitled "**Cancer Mortality Among French Atomic Energy Commission Workers**". Included in the report was a table of standard mortality ratios (SMR's) for 28 cohorts of nuclear energy workers throughout the world. The SMR is the ratio between the death rate of the group of workers (cohort) and the death rate of the public. You will notice that nearly all of these

SMR's are less than 1.00. For example, an SMR of 0.59 means that the death rate of the group was only 59% of that of the public or 41% lower. I have modified this table by adding a column labeled "Deficit (%)" for both all causes of death and cancer deaths which shows how much lower the death rate is than that of the public (expressed as a percent).

TABLE 3a. SMR for All Cause and All Cancer Reported in Cohorts of Nuclear Industry Workers

Cohort	Population Size	All Causes SMR	Deficit(%)	All Cancers SMR	Deficit (%)
NDRC	206,620	0.59*	41	0.69*	31
NRRW	124,743	0.82*	18	0.82*	18
CEA(men)	44,488	0.58*	42	0.66*	34
AEA	39,718	0.78*	22	0.80*	20
AWE	22.552	0.77*	23	0.82*	18
Springfield	19,454	0.88*	12	0.88*	12
Sellafield	14,319	0.98		0.95*	5
CEA(women)	13,535	0.74*	26	0.93	
Capenhurst	12,540	0.90*	10	0.96	
Oak Ridge	98,471	0.98*	2	0.95*	5
Hanford	44,154	0.82*	18	0.86*	14
Portsmouth	24,545	0.89*	11	0.94*	6
Los Alamos	15,727	0.63*	37	0.64*	36
Savannah River	9,860	0.75*	25	0.74*	26
Mound	4.402	0.93*	7	1.00	
Rocketdyne Al	4,563	0.68*	32	0.65*	35
CIEMAT	4,122	0.65*	35	0.77*	23
Fernhald	4,014	0.84*	16	1.09	
Savannah River	2,745	0.64*	36	0.68*	32
Mallinckrodt	2,514	0.90*	10	1.05	
AECL	8,977	0.77*	23	0.87*	13
ORNL(X10)	8,313	0.74*	26	0.79*	21
Oak RidgeY12	7,664	0.88*	12	1.00	
Rocky Flats	5,413	0.62*	38	0.71*	29
UNC	3,512	0.82*	18	0.85	
Linde	995	1.18*	-18	1.06	
CEA-DTECH	356	0.46*	54	0.77	
LLNL(SIR)		NA		1.09	
Average for all Cohorts with a (5%) significance		0.778	**22.2%**	0.790	**21.0%**

Some of the SMR's do not have enough data to be statistically significant. Those with only 5% variation are marked with an asterisk(*). I have taken the average "deficit" for cohorts with a 5% significance as shown on the bottom line. **I think that it is very important to see that the radiation workers have <u>much lower death rates</u> than the public both for all causes of death and for cancer deaths.**

31

Now let's use our common sense to understand what this means. The LNT model would have us expect a higher cancer death rate from people who were exposed regularly to extra radiation. Instead, we find just the opposite, a 21% lower cancer death rate. This is a much larger effect than the trivial amount of increased risk calculated from the LNT. Those scientists who still want to believe that the LNT model is a true picture of radiation effects have claimed that this reduction in cancer deaths is because of the "Healthy Worker Effect" (HWE).

The Healthy Worker Effect (HWE)

What is the Healthy Worker Effect? When workers are first hired, those who are healthiest are selected over others. Thus the work force starts out being more healthy than the public. Then, most workers have access to better health care than the public. The result of these factors is less sickness among the workers as well as a lower death rate. This 'Healthy Worker Effect' is real and could have an influence on the SMR's. However, I have never known of an employer who could look at a prospective worker and determine his chances of getting cancer. Cancer doesn't have such obvious early warning signs. So, any influence of the HWE on cancer death rates would only be because of the better health care available to the workers. Is this enough to cause a 21% lower cancer death rate? Better health care may lower some death rates but how much effect can it have on cancer?

In June 1988, The Industrial Disease Standards Panel, an agency of the Government of Ontario,

Canada issued a report on the Healthy Worker Effect. Several papers from experts in the field were presented. The following quotes from some of the papers are revealing.

> *"In regard to cancer, I feel that in general the HWE can be ignored in analyses that use mortality rates (SMRs)"*. Richard R. Monson

> *I conclude that the healthy worker effect is a real phenomenon, but that it is irrelevant to the interpretation of SMRs for cancer in occupational studies, so long as the first five years' observations after recruitment to the study are excluded.* Sir Richard Doll

The HWE might be ignored in our consideration of the SMRs for the Nuclear Industries Workers world wide. **This leaves us without an explanation of why the radiation workers have consistently fewer cancer deaths than the public unless we are willing to accept Dr. Cohen's view that the body's defense mechanisms, triggered by the radiation, cause a lowering of the cancer death rate.**

Just to make sure we are right in this conclusion, let's look at a special study which completely rules out the HWE. This is the Nuclear Shipyard Worker Study which was done by Johns Hopkins University under a contract with the Department of Energy and was designed to eliminate the HWE.

In this study, workers who worked on nuclear ships were compared with workers who worked on non-nuclear ships. The nuclear workers were age matched with non-nuclear workers and the mortality rates compared. The "reduction" is found by dividing the nuclear worker SMR by the non-nuclear worker

SMR, expressing this as a per\cent and then subtracting this value from 100%.

Note that the nuclear workers had significant reductions in both the total mortality and the cancer mortality.

Cohort Shipyard Workers	Total SMR	Total Reduction	Cancer SMR	Cancer Reduction
27,872 Nuclear (Hi)	0.77	**24.5%**	0.95	**15.2%**
10,348 Nuclear (Lo)	0.83	**18.6%**	0.96	**14.3%**
32,510 Non-nuclear	1.02		1.12	

Table 3b **Nuclear Shipyard Worker Study**

The LNT requires that the nuclear workers have a greater mortality rate than they would have had without exposure to radiation. But, when nuclear shipyard workers are compared to non-nuclear shipyard workers, they still have about 15% less cancer mortality, so the hormesis model fits the data much better than the LNT model.

If the radiation exposure is the cause of a lower cancer mortality, then the defense mechanisms which are triggered by the radiation may also reduce the mortality from other causes. This is exactly what appears to be the case here. The total mortality is reduced even more than the cancer mortality.

I maintain that the low level radiation has reduced the amount of cancer in the nuclear shipyard workers and in the nuclear workers world-wide. It has also reduced the mortality rate for other causes.

Additional Information about Radiation Hormesis

Some of the most interesting work on radiation hormesis was done by Dr. T D Luckey who wrote a book with the title "Radiation Hormesis"(CRC press 1991). After reviewing over one thousand studies of radiation effects on animals, he observed that over 98% of the studies support the hormesis model. After extensive study, he developed a curve showing a maximum beneficial effect at about 100 mSv/yr. (You can find other interesting information about radiation hormesis at: http://giriweb.com/luckey.htm)

The curve plots the radiation dose in units of mGy/Y. For x and gamma rays 1 mGy/Y is identical to 1 mSv/yr.

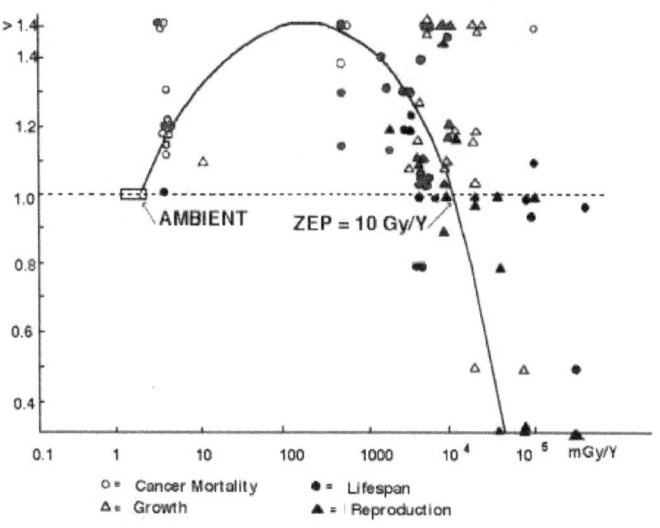

Dr. Luckey's Radiation Hormesis Plot

What Dr. Luckey calls the ZEP is the "zero equivalent point" and is the point at which the net effect of the radiation is the same as if zero excess radiation had been received. This point is at about 10,000 mGy/y or 10,000 mSv per year (1,000,000mrem/yr).

What about Radon?

Radon is a radioactive gas which emanates from the soil. All soil contains some radium-226 which decays into radon-222. Rn-222 decays with a half-life of 3.8 days into some short lived daughter products with half-lives ranging from less than a second to 27 minutes. When we breathe in radon and its daughters, a lot of the radon daughters remain and decay in the lungs. The Environmental Protection Agency has estimated that about 20,000 lung cancer deaths each year are due to radon. They base this estimate on excess lung cancer in uranium miners who were exposed to extremely high radon concentrations. They use the Linear No Threshold (LNT) model and calculate a number by multiplying very small radon exposures by many millions of people. This practice of multiplying very small doses by very large numbers of persons is discouraged by the Health Physics Society and is not reliable. Even in the range of doses where it is reliable, the LNT does not calculate probable cancers. It calculates an upper limit to the number of excess cancers. In other words, the actual number predicted by the LNT could be anywhere between zero and the calculated number.

To get a true picture, we need actual data, such as actual radon measurements and actual lung cancer mortality. Then we can see how this data compares to the theoretical model.

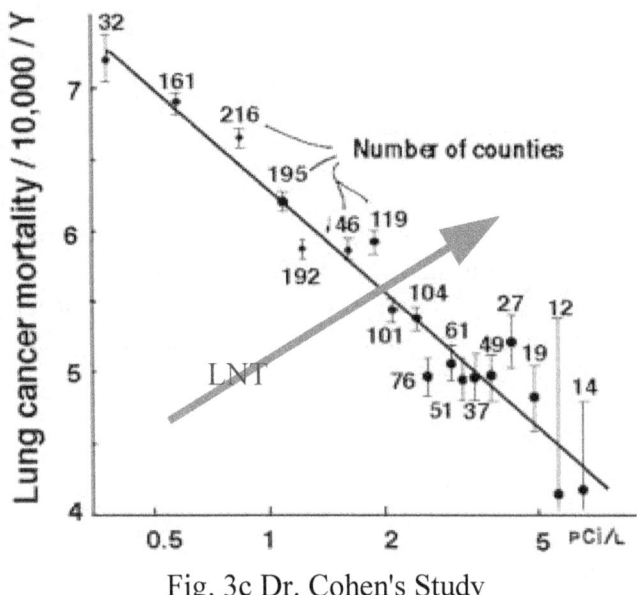

Fig. 3c Dr. Cohen's Study

After thousands of measurements of radon concentrations in homes had been made, Dr. Bernard L. Cohen made a correlation study to see how the radon concentrations compared to the lung cancer death rates. He plotted the average of all residential radon measurements in a county against the lung cancer mortality for that county. This was only one dot on the plot. He continued until he had plotted the results for over 1600 counties. He was surprised at the result shown here in Fig. 3c.

What Dr. Cohen had expected was a trend in agreement with the LNT. The arrow in Fig. 3c represents the LNT expectation with the lung cancer mortality increasing with increasing radon concentration. What he found was exactly the opposite: the lung cancer mortality decreased with increasing radon concentration. Fig. 3c includes data

37

for all the residential radon measurements in over 1600 counties. Dr. Cohen tried to find a confounding factor which could explain the results and still be in agreement with the LNT. After trying dozens of factors without success, he asked for help from others. Others tried for over 20 years without success. The data is real. The only way to explain the negative slope (decreasing lung cancer mortality with increasing radon concentration) seems to be that radon actually reduces the lung cancer mortality. Evidently, the LNT model does not work for radon at these levels and EPA's estimates are in error.

Discussion and Summary

In this chapter, we have looked at the two models for the health effects of radiation. We have found that the LNT model has serious shortcomings in explaining the effects at doses lower than 4,000 mrem (40 mSv). We have found that the Radiation Hormesis model conforms more closely to the data in this low dose region. We also found that the Hormesis model fits the mortality rates for the largest group of regularly exposed workers, the radiation workers. Their cancer death rate is dramatically less than can be explained by the LNT model and appealing to the Healthy Worker Effect. We found also that shipyard workers who were exposed to radiation had a significantly lower cancer death rate than the workers who were not so exposed. Since this study compared workers to workers, there was no Healthy Worker Effect. We also found that residential radon does not increase the lung cancer mortality but appears to lower the lung cancer mortality. This information should be sufficient to prove the Radiation Hormesis model but it is still not accepted by the official regulatory agencies. They choose to cling to the LNT model contrary to all the evidence which disproves it for lower radiation doses. Do they <u>want</u> the public to believe low doses of radiation are dangerous to health?

It may be enlightening to see just how much cancer the LNT model actually would predict for a person who received 5,000 mrem (50 mSv). The calculation is as follows:

> The NRC used 2 extra cancer deaths per 10,000 person-rem (10,000,000 person-mrem). One person receiving 5,000 mrem = **5,000 person-mrem.**
>
> 5,000 person-mrem x 2 cancer deaths divided by 10,000,000 person-mrem = **0.001 cancer death**.
>
> This would mean that his chance of dying from cancer increases by no more than 1 extra chance per thousand. (The LNT calculates an upper limit.)
>
> But the normal chance of dying of cancer is 23% or 230 chances per thousand.
>
> The extra chance would make it 231 chances per thousand or 23.1%. This change is too small to be observed.

It may be interesting to note that the cancer death statistics vary from year to year. Although the cancer deaths are currently about 23% of all deaths, the rate varied by as much as 3.2% over the 26 year period shown in chapter 4 (Fig. 4a).

If 5,000 mrem could increase the risk of cancer death from 23% to 23.1% then, according to the LNT model, 5 mrem would increase the risk of cancer death from 23% to 23.0001%. This is clearly impossible to observe. This is nothing to get excited about even if the LNT were true.

Why Worry?

Now, does it make any sense to worry about being harmed by doses of radiation which are much less than 5,000 mrem? The LNT, while claiming that all radiation increases the risk of cancer, fails to tell you that the so called increased risk is not significant. Why would you want to worry? No one gets an extra 5,000 mrem and experts agree that there is no convincing evidence of increased risk of cancer for doses lower than 5,000 mrem.

And did you know that worry has a negative health effect?

Instead of worrying, why not be optimistic and expect that the small amounts of radiation you receive could actually reduce your risk of cancer? After all, that is what happens to radiation workers. That point of view would be so much more optimistic. It would make people a whole lot happier. After all, the health effects of such small amounts of radiation, if they even exist, are too small to see.

Chapter 4: **Cancer**

You need to know how prevalent cancer is and the causes of cancer. You should be aware that radiation is not a significant cause of cancer. It takes a lot of radiation to increase the risk of cancer—many thousands of times as much as you could possibly get from a nuclear facility.

An Example of Needless Worry

While I was working with the the Utah Bureau of Radiation Control, a woman called and said she had observed several cancer cases in her neighborhood. She asked me to check her neighborhood for radiation. To appease her, I checked and found only the natural background radiation. She had assumed that, because of several cancers in her neighborhood, she was at an increased risk of getting cancer and that radiation was the cause.

How many people think like this woman and worry about being harmed by radiation? How many, when they hear of cancer, immediately think of "radiation"? If this book can eliminate some of this needless worry, it will be worthwhile.

Cancer is one of the leading causes of death. In the United States, cancer deaths currently average 23 per cent of all deaths. But this is only an average. In some communities, cancer deaths may be much more abundant – while in other communities, they may be rare. There are many known causes of cancer. Among these causes, radiation is listed. However, radiation is not a prominent cause of cancer. As we discussed previously, it takes a lot of radiation to cause an increased risk of cancer. There is no convincing evidence that a dose of less than 5,000 mrem (50 mSv) causes an increased risk of cancer.

Some Statistics

The following graph shows how the cancer death rate varied with age over a 26 year period.

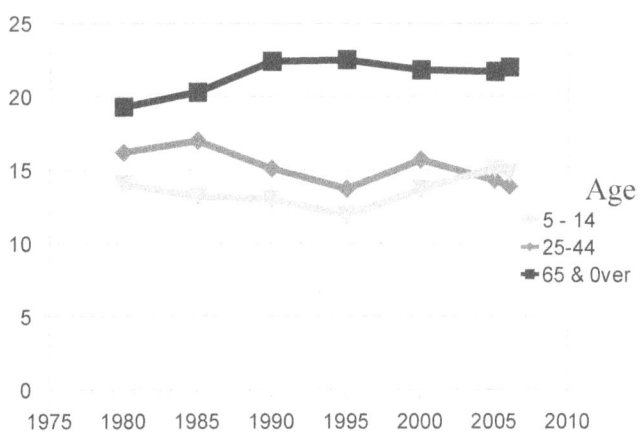

Fig. 4a US Cancer Deaths % of Total Deaths

Here is some specific information for 1999 and 2006.

1999	Number	Percent of Total Deaths	Rate Per 100,000
US	544,838	23.0	201.6
Utah	2,393	19.8	130.8
2006			
US	559,888	23.1	187.0
Utah	2,615	19.0	102.5

From this, we can see that the cancer death rate varies also with where we live. It is much lower in Utah. Life style is also an important factor in determining the risk of cancer. Besides radiation, the Mayo Clinic lists the following factors which influence the risk of cancer:

> Tobacco use.
> Alcohol use.
> Diet.
> Physical activity.
> Sun and ultra-violet exposure.
> Chemical carcinogens.
> lead, and arsenic.
> Asbestos, benzene, and formaldehyde.

These factors have a greater effect on cancer rates than low doses of radiation.

Chapter 5: **Risks from Fukushima**

You deserve to know if you should worry about small amounts of radioactivity coming from accidents like Chernobyl and Fukushima.

Could any harm come to people in the USA from the leaking reactors in Japan? The answer is No. There are many reasons why no harm could come from these leaking reactors.

In the first place, you should realize after reading the preceding chapters that it takes a lot of radiation to cause observable harm. The only way people in the USA could receive any radiation from the leaking reactors is through the air, water, or food they take into their bodies. Even though scientists may measure a small amount of radioactivity in samples of air, food, and water, you must realize that it takes billions of times these small detectable amounts to cause any effect on our health. Our instruments are very sensitive and our bodies are not very sensitive.

How Small Are Molecules?

I made a calculation to determine the chance of breathing one of the molecules of air which were exhaled by Julius Caesar in his last breath. It turns out that, if all the molecules of that air were mixed thoroughly with all the air in the entire earth's atmosphere, every breath you take should contain one or two of these original molecules from Caesar's last breath. So, you see, it is quite reasonable to expect a few of the molecules from the leaking reactors to be found in a sample of air which is larger than a thousand breaths. Just don't let it frighten you.

We cannot be harmed by these trace amounts of radioactivity coming from the Fukushima reactors. We know this since it takes radiation doses larger than 5,000 mrem (50 mSv) to cause a significant increase in cancer risk. This is over 100 times the amount of radiation we receive from nature every year.

From my analysis, I expect that no harmful health effects will come to anyone.

Let me express the reasons which I have for believing that no one in Japan will have harmful effects from the radiations from the Fukushima reactor leaks.

I know that scientists have agreed that there is no significant evidence of increased cancer risk for radiation doses less than 5,000 to 10,000 mrem (50 to 100 mSv).

Even at this large radiation dose of 10,000 mrem (100 mSv) do you know how much increase in cancer this LNT model would predict? I calculated this number and it turns out that the predicted increase in risk is a small fraction of one percent of the normal risk of dying from cancer.

For those who are interested, the calculation is shown here.

Calculation

I chose the risk factor used by the Nuclear Regulatory Commission to estimate the number of excess cancer deaths which could result from the Chernobyl accident. They chose a factor of 2 excess cancer deaths for each

10,000 person-rem (10,000,000 person-mrem). Using this risk factor, they calculated 15,000 excess cancer deaths from Chernobyl. After over 20 years, this estimate has been officially revised downward to 5,000. A dose of 10,000 mrem to a single individual would be 10,000 person-mrem. Since this is only one one-thousandth of the dose calculated to cause 2 excess cancer deaths, the risk is calculated to be 0.002 for this individual.

But about 23% of all deaths in the US are from cancer. This means that the natural risk of dying from cancer is 23 out of 100 or 230 out of 1,000. If we add to this an increased risk of 2 out of 1,000, that gives us 232 out of 1,000. It is impossible to detect an increase in the risk of death from cancer of this small amount.

We need to remember that the NRC revised their estimate for excess deaths from the Chernobyl accident down to one third of the original calculation. Therefore, we would expect the LNT prediction to be even less than 1 out of 1,000 increased risk of death from cancer. This is in good agreement with the statement by the Health Physics Society that *"below 5–10 rem (5,000 to 10,000 mrem), risks of health effects are either too small to be observed or are nonexistent."* It would seem that, according to the LNT model, it would take considerably more than 10,000 mrem (100 mSv) to produce an observable increase in cancer risk if such an increase even exists.

The following figure is a graphical representation showing how an increased risk from 10,000 mrem (100 mSv) might influence the total cancer death risk.

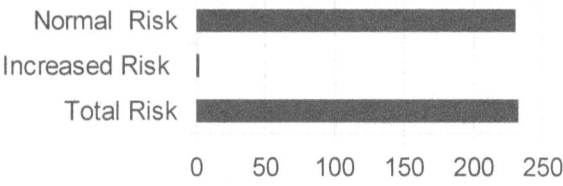

Fig. 5.1 Risk of Cancer Death per 1,000

It seems then that, in order to expect an observable increase in cancer risk, one would have to receive much more than 10,000 mrem (100 mSv) of radiation.

What is the chance that any citizen of Japan would receive this much radiation from the Fukushima reactors? Very little, but this must be analyzed.

What rate of radiation would give 10,000 mrem in one year? There are 8,760 hours in one year, so it would take 1.14 mrem (11.4 μSv) per hour. It takes 1,000 μSv to equal 1 mSv.

Reports of early radiation readings throughout the region of interest showed most readings were less than 0.50 μSv per hour with the largest reading outside the reactor buildings of 5.0 μSv per hour. Five μSv per hour would be about 44 mSv per year and much too small to cause an observable increase in cancer risk. The actual readings made on April 13, 2011 are listed here:

Tokyo Perfecture -	0.07 μSv/hr
Chiba Perfecture -	0.05 μSv/hr
Saitama Perfecture –	0.06 μSv/hr
Ibaraki Perfecture –	0.30 μSv/hr
Tochigi Perfecture –	0.30 μSv/hr
Gunma Perfecture –	0.05 μSv/hr
Yamagata Perfecture –	0.08 μSv/hr
Miyagi Perfecture –	0.20 μSv/hr

48

Iwate Perfecture –   0.04 μSv/hr
Fukushima Perfecture – 5.0 μSv/hr, 2.0 μSv/hr, 1.5 μSv/hr
Fukushima plant to about 600 μSv/hr

[Note: 1.0 mSv = 1,000 μSv]

Note: The New York Times reported on August 21, 2011 that the results of a new survey by the Ministry of Science and Education revealed a hot spot of radiation within 12 miles of the Fukushima Daiichi plant where a person who remained there for a year could receive 508.1 mSv. It also stated that the survey had found three dozen spots up to 12 miles from the plant which were above 20 mSv per year.

So, it looks like the only way a person could receive enough radiation to cause an observable increase in cancer risk is to stay at the "hot spot" found by the Ministry of Science and Education for most of a year. Even with 500 mSv per year, the LNT model would only predict an increase of 1% or change the risk from 23% to 24%.

For your information, those who believe in radiation hormesis and trust Dr. Luckey's curve (see page 31) this is about the maximum beneficial dose. Some very reliable scientists would welcome such a dose rate. Some of these scientists are actively looking for a way to increase their radiation exposure. They call it vitamin "R".

What we have been discussing is "external exposure" and the total exposure must include "internal exposure" coming from ingesting contaminated food or water. There are many measurements being taken of the food and water and the results are reported in terms of a "safe" limit. Nearly all of these readings which I have seen are well within this "safe" limit.

I believe that none of the residents of Japan will receive even close to 50 mSv in a year. Thus no significant health effects are expected.

The Emergency Workers

Now let's evaluate the greatest harm that could come to those with the greatest exposure – the emergency workers who are working to bring the reactors under control.

They have been given an emergency exposure limit of 250 mSv (25,000 mrem). If they remain within that limit, they will not experience the short term radiation effects of radiation sickness or death. The increased risk of dying from cancer which the LNT predicts from 250 mSv (25,000 mrem) is only 5 per 1,000 while the normal cancer death rate is 230 per 1,000. This is still a small increase and we have good reasons to believe that there could even be a lowering of the cancer risk if we trust Dr. Luckey's research on "radiation hormesis".

The worst possible case for the emergency workers is the harmful effect predicted by the LNT of a small increased risk of cancer which could develop within possibly 25 to 50 years. However, this is still not enough radiation to cause short term effects.

If the workers receive the limiting exposure of 250 mSv, what would this increased risk be? We know that the LNT model cannot be reliable in predicting radiation effects of doses less than 400 mSv. But, let me calculate for you what the model would predict.

> Using the risk estimator of 2 excess cancer deaths for each 10,000 person-rem which translates to 2 excess cancer deaths for each 100,000 person-mSv, and a dose of 250 person-mSv, we get:
>
> $2 \times 250/100,000 = 5/1,000 = .005$ or .5%.
>
> We also know that approximately 23% of all deaths are from cancer (US statistics).
>
> We also know that the LNT does not predict cancer deaths but only an <u>upper limit</u> to the risk of cancer deaths. The actual risk could be zero according to the LNT theory.
>
> Therefore the LNT model predicts a change in the risk of cancer death from 23% to a number in the range of 23% to 23.5%.

If the LNT model were reliable, the increased risk of cancer death which it predicts is not great, in fact, it would be too small to be observed.

Another Possibility

Now, let us look at another possibility. It is rumored that TEPCO (the Tokyo Electric Power Company) is using older, untrained persons to do this work and that they are not even monitoring their doses. If this is true, an emergency worker could exceed the emergency limit without knowing it. There is a slight possibility that he could even get "radiation sickness". However, this is not probable since the Japanese authorities have reported that only 21 of the emergency workers received more than 10,000 mrem (100 mSv) and that only 2 workers had received between 20,000 and 25,000 mrem (200 and 250 mSv). Any increased cancer risk for these older

51

workers would be meaningless since these workers are not expected to live long enough for the cancers to develop.

Lessons from Chernobyl

The information which I found on acute radiation syndrome (ARS), which is the official name for "radiation sickness", was for radiation exposures received in a short time, a few minutes to a few hours. The cleanup workers at the Chernobyl plant who were diagnosed with ARS probably received their dose within a day or two. For the workers cleaning up the reactors at Fukushima the dose rates should be much lower than at Chernobyl.

Without radiation monitoring, the first indication that an emergency worker had exceeded the limit would probably be the onset of radiation sickness which happens at about 1,000 mSv or 4 times the emergency dose limit. There is no risk of death from this amount of radiation. At the time a worker experienced radiation sickness, the worker would experience serious nausea and vomiting and not be able to continue working. Thus, he should not continue to receive additional radiation exposure. The sickness should not last for more than 60 days.

To estimate what he might expect after his recovery, we can refer to what we learned from the Chernobyl explosion.

The following quote is from "The Chernobyl Disaster and How It Has Been Understood" by Zbigniew Jaworowski.

> "From among 134 persons with this disease who had been exposed to extremely high radiation doses, 31 died soon after the accident. Among the 103 survivors, 19 died before 2004. Most of these deaths were due to such disorders as lung gangrene, coronary heart disease, tuberculosis, liver cirrhosis, fat embolism and other conditions that can hardly be defined as caused by ionizing radiation. "

This results in a death rate (among the 103 survivors of ARS) of 10.9 per year per 1,000 compared to the 2001 death rate in Russia of 13.85 per 1,000. So, the death rate does not seem to be increased. The average death rate for Russia, Belarus, and Ukraine is 15 per 1,000. The death rate among the survivors of ARS is just 73% of this average.

Which point of view is correct?

The point of view which seems to be almost universal is that there will be a large number of excess cancer deaths resulting from this catastrophic event. These numbers are only calculations using a model which has not been proved (LNT). Knowledge of the basic facts about radiation effects should remove the mystery and the worry.

The death rate following the Chernobyl accident has not increased as predicted by the LNT model. Instead, it seems to be lower! The following quote from "The Real Chernobyl Folly" by Zbigniew Jaworowski will illustrate this.

> "Data published by Ivanov et al. (2004) and cited in the Chernobyl Forum documents (Forum 2005; Forum 2006) show a 15 to 30 percent lower mortality among the Chernobyl emergency workers, and a 5 percent lower

average solid cancer incidence among the people in the Bryansk district (the most contaminated area in Russia) in comparison with the general Russian population."

So, there is very little chance that an emergency worker would get enough radiation to cause "radiation sickness" even if he is not monitored. And even in the worst case, a worker who received enough radiation to have "radiation sickness" should survive and recover after being sick for a few months at most. He should not expect an increased risk of death, and I would not expect any cases of radiation sickness.

The views which I have expressed are based on the following facts:

- There is no conclusive evidence that radiation doses less than 5,000 to 10,000 mrem (50 to 100 mSv) increase the risk of cancer.

- Survivors of the Chernobyl accident who were diagnosed with radiation sickness and survived did not have an increased death rate.

The most harmful effect from the damaged nuclear reactors will be from the <u>fear</u> that misinformation about radiation effects has fostered. <u>That was the only really harmful radiation effect on the people of Europe following the Chernobyl explosion.</u>

Summary

This analysis is based mainly on the fact that scientists agree that radiation exposures less than 5,000 to 10,000 mrem (50 to 100 mSv) have not been shown to cause harmful effects. All estimates of harmful effects from individual exposures less than this are simply calculations using a model which has not been proven and thus are unreliable.

It is also important to know that the hormesis model, which predicts beneficial effects from low doses of radiation, is supported by a large amount of scientific evidence.

Please understand that the expectation of significant radiation injury and cancer is due to a misunderstanding of the true health effects of radiation. If we understand the radiation effects, we will not be worried about being harmed by the reactor leaks.

In the United States, we receive much more natural background radiation than we could ever get from the radioactive material escaping from the Japanese reactors and dispersed throughout the atmosphere on its way to the US. This "man-made" radiation cannot be distinguished from "natural" radiation.

Chapter 6: How Dangerous is Plutonium?

Rumors

You may have heard some of these rumors.

- Plutonium is "the most toxic substance known to man".

- One pound of plutonium could kill 8 billion people.

- A single particle of plutonium inhaled into the lung can cause cancer.

These rumors are not true but, if you have heard these rumors and believe them, you could be very worried about any plutonium escaping from the Japanese reactors.

I believe that the public deserves to know the truth about plutonium.

Dr. Bernard L. Cohen, a radiation scientist, has analyzed the danger of plutonium and determined that its toxicity is about the same as caffeine. In fact, he offered Ralph Nader the challenge that he would eat as much plutonium as Mr. Nader would eat caffeine on public television. Mr. Nader did not accept the challenge. Obviously Dr. Cohen does not believe that plutonium is as toxic as has been claimed.

Just how toxic is plutonium? To dispel the rumors, let us look first at the **theoretical dangers** and then at the **actual experience** of people who have inhaled or ingested significant amounts of plutonium.

Theoretical Dangers

The perceived danger of plutonium is from its radioactive properties. Pu-239 decays by emitting a 5.1 MeV alpha and some very weak betas and x-rays. The entire danger is from the alpha radiation. Since alpha rays cannot penetrate the dead layer of skin, the alpha emitter must get inside the body to be any danger. If Pu-239 is ingested only about 0.05% is absorbed into the blood stream and 99.95% is eliminated. If the Pu-239 is inhaled, it is about 225 times as dangerous.

Dr. Bernard L. Cohen, a radiation scientist, has made a careful analysis of the life-time cancer death risk from inhalation of plutonium and estimates it to be about 25 fatalities per pound of plutonium dispersed in the air. He made this calculation using the BEIR method so it includes the linear-no-threshold (LNT) hypothesis.

A "Human Health Fact Sheet" on plutonium by Argonne National Laboratory gives the life-time risk of dying from cancer from inhaling 1 pCi (piccocurie) of Pu-239 as 0.029 per million. The normal risk of dying from cancer is about 230,000 per million or about 7.9 million times as great.

Plutonium vs. Radon

Since the risk is from alpha radiation and we breathe air containing radon (which is an alpha emitter), I have compared the Pu-239 radiation to the radiations from radon and its daughters (decay products) in the following chart.

		Radiation Energy (MeV)		
	Half-life	α	β	γ
Pu-239	24,000 years	5.1	<	<
Rn-222	3.8 days	5.5	<	<
Po-218	3.1 min	6.0	<	<
Pb-214	27 min	--	0.29	0.25
Bi-214	20 min	--	0.66	1.5
Po-214	160 μsec	7.7	<	<

Notice that the Pu-239 has only the one alpha while the Rn-222 and its daughters with half-lives less than 30 minutes have 3 alpha's totaling 19.2 MeV. There are other daughters of Rn-222.

Pu-239 decays to U-235 with a half-life of about 710,000,000 years and decays so slowly that none of its radiation will be observed.

Since radon is a gas with a half-life of 3.8 days, the radon inhaled will be exhaled and only its short half-lived daughters, which are not gases, will remain to decay in the lung. Plutonium is not a gas and all the plutonium inhaled can remain in the lung to decay.

The radon daughters attach to dust particles in the air and their activity should be at least 25% of the radon activity. Considering this and the nearly 4 times the alpha energy as the plutonium alphas, 1 pCi of radon releases about the same alpha energy to the lung as 2 pCi of plutonium. Therefore, a home with the EPA recommended maximum of 4 pCi radon per liter of air should give the lungs as much alpha energy as 8 pCi of Pu-239. So plutonium is not as "deadly" as radon.

Actual Experience

All these speculations and calculations are just theoretical. What we need are some factual data about the experience of individuals who have been exposed to significant amounts of plutonium. I have been able to find some information about such individuals.

Early Workers Exposed at Los Alamos

26 male workers were exposed to plutonium in 1944-45. They were estimated to have had plutonium depositions of from 2 to 95 nanocuries (nCi) with an average of 26 nCi.

[One pound of Pu-239 = 28 billion nCi so 26 nCi is about one one-billionth of a pound.]

After 37 years, there had been only 3 deaths among these 26 workers. One was by heart attack, one by congestive heart failure and one was by accident. For this group there would have been 6.6 expected deaths based on U.S. adjusted rates for white males.

Thus there is no evidence of any harm to these workers from the plutonium. There were no cancers.

Later Los Alamos Workers

224 white male workers at Los Alamos were identified to have 10 nCi of plutonium or more as of 1974. A study of these workers reported in 1983 that 43 workers had died from all causes compared to an

expected 77 deaths based on U.S. adjusted rates. This is only 56% of the expected rate.

It appears that the plutonium was not causing deaths in these workers.

Rocky Flats Workers

A mortality study of Pu and other workers at the Rocky Flats nuclear facility by Wilkinson, et al. in 1983 observed 334 deaths from all causes for these white males compared to an expected 522 based on U.S. adjusted rates. (64% of the expected rate.) Deaths from all cancers were 79 observed compared to 105 expected. (75% of the expected rate.)

It appears that the plutonium was not causing excess deaths or excess cancers in these workers.

The information for the above studies was taken from "A 37-Year Medical Follow-Up of Manhattan Project Pu Workers" by George L. Voelz and Robert S. Grier

(Health Physics Vol. 48, No.3)

A Three Laboratory Study

A study of three National Laboratories handling plutonium showed that workers who were exposed through inhalation and had a body burden of plutonium in excess of 2 nanocuries (74 Bq) had consistently and significantly lower lung cancer mortality than the public. In the following table, a ratio of 1.0 would indicate a rate equal to that of the public.

PLUTONIUM AND LUNG CANCER
Gary L. Tietjen (Health Physics, May 1987)

Laboratory	Workers	Observed / Expected
Hanford	84	.29
Rocky Flats	386	.14 (.21)*
Los Alamos	130	.20

> *Ratios were compared with the public except that the Rocky Flats employees were also compared to fellow workers. This is the number in parentheses.

From this study, the workers with a body burden of plutonium had only about one fourth as many lung cancer deaths as their fellow workers who were not so exposed.

These results demonstrate that plutonium in any amount does not always cause cancer as some news media persons have claimed. In fact, it demonstrates that substantial amounts of plutonium can be inhaled with no resulting cancers. There could even be a protective effect reducing the risk of cancer.

(You will remember from Chapter 3 that there is evidence for radon having a protective effect against lung cancer and radon emits radiation similar to that from plutonium.)

The above examples are for people with relatively small plutonium burdens.

Plutonium workers at Hanford, Los Alamos, and Rocky Flats (in the United States), and Sellafield (in the United Kingdom) had body burdens of Plutonium less than 1,000 Bq while workers at the Mayak

Nuclear Complex in Russia had mean body burdens over 1,000 Bq and as high as 9,200 Bq with individual uptakes up to 470,000 Bq. The Mayak workers were the only ones showing evidence for an association between cancer mortality in the lung, liver, and bone and the uptake of plutonium.

This information was taken from "Cancer Mortality Risk among Workers at the Mayak Nuclear Complex " by N. S. Shilnikova et. al. Published in RADIATION RESEARCH 159, 787–798 (2003)

Summary

In summary, I have quoted three claims about plutonium toxicity.

1. The claim that plutonium is "the most toxic substance known to man". This is false. Even caffeine which is taken every day by millions is as toxic as plutonium.

2. The claim that "One pound of plutonium could kill 8 billion people" is false. Dr. Cohen calculated that one pound of plutonium, dispersed in the air could kill only 25 people and these would be from cancer and would take many years to develop. Besides, the calculation of 25 deaths used the LNT which only calculates an upper limit.

3. The claim that a single particle of plutonium inhaled into the lung can cause cancer is false. Experience has shown that body burdens of up to 1,000 Bq (27 nCi) of plutonium have not shown increased cancer.

These rumors are kept alive by the news media. They get more attention for an article about plutonium if the public believes it is extremely dangerous. The articles always seem to say "deadly" plutonium or "deadly" radiation. For example: "Deadly Plutonium found in Soil at Japan's Fukushima Nuclear Plant".

Actual experience has shown that workers with plutonium uptakes of less than 1,000 Bq had significantly lower cancer death rates than the public.

Plutonium is not "deadly". Without factual information the public tends to panic when trace amounts of plutonium are reported in the soil or in air filters. This is not justified by science.

Chapter 7: **What You Should Have Learned From This Book**

- We receive a lot of radiation from our natural background. In Utah it averages about 80 mrem per year. The average for the United States is about 40 mrem per year.

- A person will receive less than 1 mrem in a visit to a nuclear facility.

- It takes a lot of radiation to cause a harmful health effect.

  - Over 10,000 mrem to possibly increase the cancer rate.

  - 100,000 mrem in a short time to cause radiation sickness.

  - 500,000 mrem in a short time to cause radiation death.

- The most commonly accepted theory of radiation health effects is the linear-no-threshold theory or LNT. The LNT has never been proven and is not reliable for doses less than 40,000 mrem (400 mSv).

- The statement "All radiation is harmful" is not true.

- Low doses of radiation reduce the amount of cancer.

- Residential radon reduces the amount of lung cancer.

- Cancer is more common than you may realize and accounts for about 23% of all deaths in the United States. It varies from year to year and from one location to another. In Utah, it accounts for about 19% of all deaths.

- Since there is no evidence of increased cancer for doses less than 50,000 to 100,000 mrem (500 to 1,000 mSv), there is no danger that the people of Japan will have increased cancer because of the Fukushima reactor leaks.

- Plutonium is not the most toxic substance known to man and is probably no more toxic than caffeine. Workers who have inhaled significant amounts of plutonium have much lower death rates than their fellow workers.

- LOW DOSES OF RADIATION ARE NOT DANGEROUS.

For more information visit

understanding-radiation.com

Appendix 1: Author's Qualifications

EDUCATION

1964 M.S. Nuclear Physics   BYU
1961 M.S. Radiological Health   NYU
1953 B.S. Math and Physics   Ricks College
Additional Study at:
    Brigham Young University
    Michigan State University
    University of Utah
    University of Rochester
Numerous short courses by the NRC, DOE, and EPA

RADIATION RELATED WORK EXPERIENCE

Health Physicist   Nuclear Reactor
Radio-physical chemist Food Irradiation Project
Radiation physicist   Regional Medical Program
Health physicist   Radiation Control Program
Radiation specialist   Radioactive Waste Facility
Private Consultation

• Certified Health Physicist (certified 1978)

• Atomic Energy Commission Graduate Fellow in Radiological Physics at the University of Rochester and Brookhaven National Laboratory 1953-54.

About the Author

Blaine Howard received a B.S. in Mathematics and Physics from Ricks College in 1953. He attended the University of Rochester in 1953 on a Graduate Fellowship in Radiological Physics from the Atomic Energy Commission.

In 1954, he was a Jr. Health Physicist at the Brookhaven National Laboratory research reactor until drafted into the U.S. Army for 2 years. In 1958, Howard worked for the University of Utah at Dugway doing high level dosimetry for the the U.S. Army's food irradiation project. In 1960, he was a research assistant at New York University under Dr. Merrill Eisenbud and received his M.S. in radiological health in 1961. After receiving an M.S. in nuclear physics from BYU in 1964, Howard taught physics at Eastern New Mexico University and at Cal Poly, San Luis Obispo.

After spending 2 years in a Ph.D. program in physics at Brigham Young University, he worked at the Intermountain Regional Medical Project as a radiation physicist.

He then spent the next 18 years as a health physicist for the State of Utah's radiation control program. During this time he served as technical advisor to Governor Matheson's High Level Waste Task Force, and as chairman of a committee to select a site for disposal of the Vitro Uranium mill tailings. He had health physics responsibility for the Vitro relocation project.

In 1990, Howard started work for Envirocare as Site Radiation Officer and as a Health Physics Specialist, he conducted a gamma spectroscopy program.

Howard is presently retired but has taught some math and physics classes at the Salt Lake Community College and has done some substitute teaching in math and science at high schools in the Jordan School District.

www.ingramcontent.com/pod-product-compliance
Lightning Source LLC
Chambersburg PA
CBHW071612170526
45166CB00003B/1069